Global Warming

&

Human Cosmic Purpose

A catalogue record for this book is available from the British Library

I.S.B.N: 978-1-907962-74-5

Publication Date: 02 02 2020

Published by Cranmore Publications

Exeter, England

cranpubs@gmail.com

"What is man? – so I might begin; how does it happen that the world contains such a thing, which ferments like a chaos or moulders like a rotten tree, and never grows to ripeness? How can Nature tolerate this sour grape among her sweet clusters?"[1]

Friedrich Hölderlin

[1] Hölderlin, F. (1797) 'Hyperion' in Eric L. Santner (ed.), Hyperion and Selected Poems, New York: Continuum, 1990, p. 35.

Dear Friedrich

The human species is the zenith of the evolutionary progression of life on Earth. That which troubles you, the chaos and the rottenness, enables the bringing into being of wondrous life-giving fruits. The need for these joyous delicacies explains the tolerance of the sourness that precedes them. Rest assured my friend, the human species will reach ripeness, but first it needs to fulfil its cosmic purpose!

'Save the planet'

I wonder how many times that you have been urged by someone – a friend, a celebrity, a politician, an environmentalist – to change the way that you live in order to help to 'save the planet'?

Such urgings are commonplace in our current stage of Solar-Systic unfoldment, the 'guilt-tinged' age of the technological explosion. Such urgings currently pervade the media, popular culture and academia. Such urgings are central to the way that we have come to see ourselves as a species. As the age of the technological explosion has advanced these urgings have escalated; we are now faced with a pandemic of urgings to 'save the planet'.

But, does the planet need to be saved?

'Save life on Earth'

The planet Earth does not need saving; those who urge us to change the way that we live know this. When they talk of the need to 'save the planet', what they really mean is that they believe that there is a need to save the life that resides on the Earth.

Some people believe that life on Earth in its totality needs to be saved; in other words, they believe that the Earth could soon become a barren and lifeless place. Other people believe that it is only part of life on Earth that needs to be saved; in other words, they believe that a certain proportion of the species/life-forms that currently exist need to be saved, but that if this proportion of life-forms were to all die that the Earth would still be home to a plethora of wonderful life-forms.

The planet Earth itself is exceptionally robust; there is no sense in which it needs to be saved.

But, does the life that resides on the Earth need to be saved?

What is it that life on Earth supposedly needs to be saved from?

Those people who urge us to 'save the planet' / 'save life on Earth' do so because they believe that life on Earth needs to be saved due to widespread perturbations and destabilisations that have been caused by the activities of the human species. They believe that these widespread perturbations and destabilisations pose a threat to the continued thriving, even the continued survival, of life on Earth.

In other words, the human species has come to conceptualise itself as an entity that poses a threat to the continued thriving of life on Earth, due to the magnitude of its collective impact on the Earth's biosphere. The current pandemic of urgings to 'save the planet', are urgings for the human species to rein in its destabilising activity, in order to avert its supposed potentially catastrophic impact on the Earth's biosphere. The ethos underlying this conceptualisation, and its associated pandemic of urgings, is clearly that the interests of life on Earth are best served by the human species 'getting out of the way' and 'leaving things to nature' as much as possible.

In short, the pandemic of urgings has arisen due to the belief that life on Earth needs to be saved from the destructive activities of the human species.

Saving ourselves

When some people urge us to 'save the planet' what they really mean is that they believe that the human species needs to be saved; they believe that the human species needs to change the way that it collectively acts in order to save itself.

To believe such a thing is not to deny that the perturbations and destabilisations caused by the human species can bring suffering and death to non-human life-forms. One can accept that this is the case whilst also believing that, fundamentally, it is the human species that needs to be saved. These people believe that life on Earth was thriving before the human species came into existence, that the perturbations and destabilisations brought into being by the human species might lead to the extinction of both the human species and some other species, but that if this eventuality came to pass that life on Earth would be fine and dandy and would be positively thriving in the distant future. According to this way of thinking, whilst the human species, and some other species, would no longer exist, they would have been replaced by other wonderfully complex and interesting life-forms and life on Earth would be flourishing.

To talk of 'saving ourselves' is, as we have been exploring, usually to talk of preventing the extinction of the human species. However, one could use the phrase in a less alarmist way, and take it to mean 'saving human civilisation in its current form'. For example, if sea levels were to spiral upwards causing most of the world's cities to become submerged into oblivion, then the human species wouldn't become extinct, but human civilisation would be very different to the way that it is now.

In short, some of the people who are urging us to change the way that we live do so because they believe that we need to save ourselves. They believe that life on Earth, as a totality, as a flourishing entity, does not need to be saved, but that the human species does need to be saved.

The trap of the flawed extrapolation

We have seen that some people believe that we need to change the way that we live in order to save ourselves, whilst also believing that life on Earth would have a long and glorious future if we were to go extinct. This isn't true; such a belief is wrong. This erroneous belief derives from a flawed extrapolation from the past to the future.

Life on Earth was thriving for millions of years before the human species came into existence, so it is very easy to fall into the trap of believing that it would continue to thrive in the future if the human species became extinct. In the past, life on Earth has encountered serious setbacks, what we call 'mass extinction' episodes. Following each of these past episodes life on Earth has bounced back to a level of health that has superseded that which existed prior to the episode.

We are in the midst of a human-induced mass extinction episode, but the past is currently not a good guide to the future in terms of whether life would bounce back and thrive following a mass extinction episode which included the extinction of the human species. Such an extinction episode, at our current stage of Solar-Systic unfoldment, would propel us down the path that leads to life on Earth being forever decimated. If one believes that following such an episode life would bounce back and thrive, like it has in the past, then one has fallen into the trap of the flawed extrapolation.

Our mission

The human species does need to be saved; there is no doubt about that. However, our mission, throughout the rest of this book, is to come to appreciate why it is not just the human species that needs to be saved; rather, it is the broader phenomenon of life on Earth that needs to be saved. When one has acquired this appreciation, then one will have escaped the clutches of the trap of the flawed extrapolation.

The phrase 'life on Earth' refers to that part of the Earth that is living, as opposed to non-living. Life on Earth can be absolutely thriving, like it is currently, being constituted out of a ginormous plethora of wonderfully complex life-forms. Life on Earth can also be in a terrible state, containing nothing more than simple single-celled organisms.

To save life on Earth is to maintain the state of thriving. If life on Earth exists in the future, but is in a terrible state, then all is lost; life on Earth would be close to lifeless and would be lurching headlong into oblivion.

Let us explore why life on Earth needs to be saved. And, let us come to understand why the cause of the need for saviour has nothing to do with the human species.

The main threat to life on Earth

There is no doubt that life on Earth needs to be saved. The cause of the need for saviour is that our Solar System has unfolded to the point at which the ability of life on Earth to continue to keep atmospheric conditions favourable for its continued thriving, through the removal of carbon dioxide from its atmosphere, has almost ceased.

Atmospheric carbon dioxide concentrations on the Earth used to be 5000 parts per million (ppm). That is not a typing mistake, I didn't mean to say five hundred ppm; there used to be five thousand ppm of carbon dioxide in the Earth's atmosphere! Where has all that carbon dioxide gone?

As our Solar System has unfolded the amount of Solar Radiation that has been propelled to the Earth from the Sun has continuously increased. This continuous increasing is a force for the warming of the Earth's atmospheric temperature. If this force wasn't offset by an opposing force, the Earth's atmospheric temperature would now be far too hot for life to be thriving. Yet, life on Earth is currently positively thriving.

Life on Earth is currently positively thriving because it has responded with an opposing force, in order to keep its atmospheric temperature favourable for its continued flourishing. This opposing force is the pulling of carbon dioxide out of the Earth's atmosphere and its storage under the Earth's surface. This is why atmospheric carbon dioxide concentrations have fallen from 5000ppm to only a few hundred ppm.

When the human species started to have a significant impact on the Earth, at the start of the Industrial Revolution, life on Earth had offset increasing incoming Solar Radiation to such an extent that the level of atmospheric carbon dioxide had fallen from 5000ppm to 280ppm. That is a colossal reduction! Through this colossal reduction life on Earth has been able to prevent its atmospheric temperature from rising to a level that is too hot for life to survive and thrive.

What does this mean for the future of life on Earth? The amount of Solar Radiation reaching the Earth is only going to keep on increasing in the future. How can life on Earth continue to thrive in the face of this continued continuous increasing? What has happened in the past clearly cannot continue. In the past, a fall from 5000ppm to 280ppm has been required to offset the increasing Solar Radiation and thereby maintain the Earth's atmospheric temperature. Atmospheric carbon dioxide cannot fall much lower than this. One can easily comprehend that, if what has been needed in the past, was to continue to work in the future, then atmospheric carbon dioxide concentrations would not only need to fall to zero, they would ultimately have to fall to minus several thousand!! That is obviously not possible!

Atmospheric carbon dioxide concentrations cannot fall much below 280ppm because the thriving of life on Earth requires a significant amount of carbon dioxide to be in the Earth's atmosphere. The existence of life requires plants to perform photosynthesis (take carbon dioxide from the atmosphere) and to respire (put carbon dioxide into the atmosphere), and the existence of animals entails putting carbon dioxide into the atmosphere via respiration. So, obviously, a thriving life-bearing planet, a planet which has trillions upon trillions of flourishing life-forms, will need to have quite a lot of carbon dioxide in its atmosphere. Following a reduction to 280ppm, from 5000ppm, there isn't much scope for atmospheric carbon dioxide levels to fall much further. If life is to continue to survive and thrive then another solution, another opposing force, is desperately needed.

This is the cause of the main threat to life on Earth. Increasing Solar Radiation has almost overwhelmed the ability of life on Earth to maintain its atmospheric temperature, to keep it down to a level that is conducive for its continued thriving, through reducing its 'greenhouse effect' via atmospheric carbon dioxide removal.

Life on Earth is in grave peril; it needs to be saved. It needs to be saved from increasing levels of incoming Solar Radiation.

How can life on Earth be saved?

Life on Earth is currently in grave danger. For hundreds of millions of years life on Earth has successfully maintained the conditions that it requires for its continued flourishing through offsetting increasing Solar Radiation by reducing atmospheric carbon dioxide concentrations. We can call such an offsetting/reducing life on Earth's 'first line of defence'.

The fall in atmospheric carbon dioxide concentrations to 280ppm means that the 'first line of defence' is speeding towards eternal impotence. If life on Earth is to continue to survive, and thrive, it requires a new way to maintain its atmospheric temperature in the face of forever increasing levels of incoming Solar Radiation.

There is only one way in which life on Earth can be saved. Life on Earth needs to develop the capability to deflect incoming Solar Radiation, thereby preventing it from reaching the Earth's atmosphere. Such deflection is life on Earth's 'main line of defence'.

The human species

The human species is that part of life on Earth which has the ability to bring forth the 'main line of defence'. In other words, the human species is the saviour of life on Earth. In the absence of the human species there would be no 'main line of defence'. And, without the possibility of a 'main line of defence' life on Earth would be lurching towards oblivion. If the human species did not exist, the day of obliteration for life on Earth would be nigh.

Thankfully, life on Earth has complexified to the point that is the human species, and it thus has the ability to bring forth the 'main line of defence'. Life on Earth, thanks to the human species, has a long and glorious future.

The 'guilt-tinged' age of the technological explosion

Given that the human species is the saviour of life on Earth, why is it that we are currently faced with a pandemic of urgings to 'save the planet' from the destructive and destabilising activities of the human species?

Bringing forth its 'main line of defence' requires life on a planet to go through a technological explosion. More specifically, a stage in Solar-Systic unfoldment is reached when, due to the inevitable impending impotence of its 'first line of defence', the epoch of the technological explosion is needed if a thriving life-bearing planet is to continue to flourish; without it, life on that planet is doomed. However, it is also an atrocious epoch; a truly ghastly period of Solar-Systic unfoldment. The bringing forth of technology entails the bringing forth of immense disruption, large destabilisations and tremendous suffering. These things come into being as the bringer forth of technology explores its planet, as it learns how to master the resources that are at its disposal, as it hones its abilities as the master modifier of its Solar System.

As technology explodes on a planet, suffering also explodes on that planet. The master modifier itself suffers immensely, as do the life-forms which the master modifier inevitably sees as resources for its use. The nature of the master modifier, the human species, is to master modify. So, the master modifier master modifies.

However, a point comes, in the midst of the technological explosion, when the master modifier realises that its master modifications have had consequences that it did not foresee. The master modifier was just acting in accordance with its nature; it didn't think about consequences; indeed, until very late in the day there were no significant consequences to be thought about. Almost in the blink of an eye the human population size exploded, technology exploded, and the master modifier suddenly had a massive disruptive and destabilising impact on the Earth. When the master modifier comes to realise that its past actions have serious consequences, when it comes face-to-face with the suffering that it has brought into being, when it comes to realise the extent to which it has destabilised the biogeochemical cycles of the Earth, then it naturally becomes overwhelmed with guilt.

There is fear, there is anguish, and there is despair. What have we done to the planet? Why are we so selfish, and greedy, and stupid? We need to stop what we are doing! We need to rein in our activities!! Such is the initial response of the master modifier when it comes to realise the extent of its impact on the Earth in the 'guilt-tinged' age of the technological explosion.

This initial response, this guilt, explains the urgings of those who want us to 'save the planet' from the destructive and destabilising activities of the human species. Those who urge us to rein in our activity in order to 'save the planet' are overwhelmed by the guilt that comes to pervade human culture in the 'guilt-tinged' age of the technological explosion.

The partial reversal of the 'first line of defence'

In the epoch of the technological explosion the master modifier rampantly utilises all of the resources that it can fruitfully make use of on its host planet. These resources include the fruits of the 'first line of defence'. In other words, the master modifier releases the fossil fuels that were previously stored underground as life on Earth successfully regulated its atmospheric temperature in the pre-human era through the pulling of carbon dioxide out of its atmosphere. The fruits of the 'first line of defence', what we call 'fossil fuels', power the technological explosion which brings forth the 'main line of defence'.

The release of the fossil fuels that were previously stored under the surface of the Earth is a partial reversal in life on Earth's 'first line of defence', as it brings forth its 'main line of defence'. In other words, the increasing level of atmospheric carbon dioxide that is occurring at the moment, due to the human release of previously stored fossil fuels, is a sign that the master modifier is in full swing; it is a sign that life on Earth is well on its way to bringing forth its 'main line of defence'.

If one was to observe the pattern of atmospheric carbon dioxide concentrations that have occurred on the Earth, a massive fall over a prolonged period (from 5000ppm to 280ppm on the Earth), followed by an abrupt spike (from 280ppm to over 400ppm on the Earth), on any life-bearing planet, then this would be a sign that all is well on that planet. It indicates that the 'first line of defence' has done its job and that life on that planet is well on its way to bringing forth its 'main line of defence'.

If one is observing the atmospheric carbon dioxide concentrations on a distantly located life-bearing planet, and there is a prolonged fall over a long period of time, but there is no sign of an abrupt spike, then one should be deeply and increasingly worried for the life on that planet. One would be praying to see such a spike soon, for without it one would know that the life that resides there would be hurtling towards extinction. In other words, one would be praying to see global warming that has been initiated by that planet's master modifier.

Environmentalism

We have seen that the realisation of the magnitude of the human impact on the Earth gives rise to a sense of guilt concerning what the human species has done. When this sense of guilt has arisen it inevitably leads to the phenomenon which has been dubbed 'environmentalism'. In the face of anger, guilt and despair, emotions are aroused, and growing numbers of humans seek to alleviate their guilt through reducing their impact on the Earth. They also seek to urge other people to do the same; and, those who do not act in accordance with these urgings become the 'enemy'. Such urgings, and the associated creation of the 'enemy', is perfectly understandable and inevitable. Such urgings play a part in bringing about a truly sustainable and peaceful world.

However, as we have seen, if life on Earth is to survive, and thrive, it requires its master modifier, the human species, to bring forth its 'main line of defence'. And, this bringing forth means that the disruptive epoch of the technological explosion needs to be passed through. So, whilst a pervasive sense of guilt emerges in the epoch of the technological explosion, the human species hasn't done anything that it should feel guilty about. The human species has been acting the way that it has in order to bring forth the 'main line of defence' so that it can save life on Earth. Saving life on Earth is a noble endeavour; it isn't something to feel guilty about!!

Of course, when it is fully enmeshed in the 'guilt-tinged' age of the technological explosion, the vast majority of the human species does not realise that its actions are contributing to the outcome of saving life on Earth, so all that is left is the guilt! Feeling such guilt is therefore natural and normal; after all, in the ghastly epoch that is the age of the technological explosion the master modifier of its planet and solar system inflicts suffering and death onto its fellow planetary life-forms on a ginormous scale.

Human suffering

The airplane crashes. Hundreds of people die. Suffering. Death. Mourning. Misery for friends and family.

The school bus gets ploughed into by a lorry. All the children die. Suffering. Death. Mourning. Misery for friends and family.

A machine gun massacre. The nuclear bomb explodes. The roller coaster comes off its rails. The bungee jump cord snaps. The explosive device detonates. The car ploughs into the pedestrians. The air of the city is full of fumes and smog created by technological society; the people who breathe the air become immensely sick.

The masses flock to live in concrete jungles and spend their time immersed in concrete and engulfed in technology; they spend most of their time staring at the screens of technological devices. Is this good for their health? Of course it isn't! They need forests, flowers, fresh air and the sea; yet, the masses flock to the concrete jungles and stare at the screens of technological devices! Such a lifestyle brings them immense mental ill-health and physical ill-health. Many cannot take it anymore; they throw themselves off the tall buildings humans have built; they shoot themselves with the guns humans have created; they shoot others with the guns humans have created; they cut their wrists with sharp knives and razor blades; they drink alcohol until they get into a state of oblivion.

The masses eat highly processed, chemical-filled foods which have been produced by mass agriculture, and which have been degraded by being zapped in microwave ovens and decimated in deep fat fryers. The masses eat animals, which their bodies were not designed to eat. The masses make themselves terribly ill through all of this. Why don't they eat fresh, organic, healthy food – fruits, vegetables and grains? Why do they eat degraded food that degrades their mental and physical health?

Why do humans live like this? Why do humans do all of this to themselves? If humans had a simple life; for example, if they transported themselves under their own steam, like other animals, using their own two feet, there would be no carnage. Currently over a million humans a year are killed in road accidents; slaughtered by motor vehicles on roads. The precious children who were in the bus that was ploughed into by the lorry would still be alive if humans transported themselves by foot!

It is not human nature to transport itself by foot. It is not human nature to live a simple technology-free life. The human is not just another animal. The human species is the master modifier of its planet, and solar system; it has no option but to act in accordance with its nature and to master modify. Transform! Modify! Create! Analyse! Investigate! Utilise! Modify some more! Invent! Invent better and faster! Modify more and more and more!!!

In the atrocious epoch that is the technological explosion the human species is bringing forth increasingly complex technologies, for the benefit of life on Earth, so that it can fulfil its cosmic purpose, but in so doing it suffers tremendously. The human species is currently in the stage of Solar-Systic unfoldment where it is unleashing technology, but it has yet to master the technologies that it is bringing forth. The technologies are running amok; consequently, humans are suffering and dying in their hordes. Millions upon millions of humans slaughtered due to the technologies that humans have brought into being!

Such is the atrocious epoch that is the age of the technological explosion. It is the extreme magnitude of the technology-induced misery, suffering and death, both for the human species, and for the rest of life on Earth, that makes this epoch so atrocious.

The future will be better. The human species will eventually both master its technologies and master itself so that it doesn't abuse the technologies that it has brought forth. The suffering of the human species will eventually be brought down to a bearable level as our Solar System continues to unfold and we move out of the atrocious epoch that is the age of the technological explosion.

The enlightened humans of the future

If we fast forward a few thousand years into the future, we can appreciate that the human species will be much more enlightened, immensely more spiritually developed. The humans alive at this time will, no doubt, look back at the epoch of the technological explosion and think to themselves: "How could our ancestors possibly have used and abused their fellow planetary life-forms to such an extent?

How could they possibly have lived with themselves? Causing such abuse must have pushed them into insanity! They must have been overcome with guilt concerning what they were doing!"

Yet, these more enlightened humans of the future will understand perfectly well why this atrocious epoch needed to be passed through; for, without it, they know that the peaceful, sustainable planet that they inhabit would never have come into existence. They know that without the 'main line of defence' life on Earth would have been obliterated. They know that if, in the epoch of the technological explosion, the human species wasn't deluded concerning its place in our Solar System, if it didn't see its fellow life-forms as inferior resources to be used, if it didn't see itself as the selfish destroyer rather than as the precious saviour, if it didn't become overwhelmed with guilt, then the master modifier would not have been able to bring forth the wonderful 'main line of defence'.

One cannot master modify, exploit, enslave, slaughter, factory farm, vivisect, eat, that which one sees as similar to oneself; the master modifier, in the epoch of the technological explosion, has to see itself as special, as superior; it needs to believe that its fellow planetary life-forms are very different to itself, when in reality they are not.

Delusion

In the age of the technological explosion the human species becomes deluded. There are three strands to this delusion. Firstly, the human species is deluded because it does not realise that it is the precious saviour of life on Earth; rather, it believes that it is fundamentally a destructive influence on the Earth, as far as life is concerned. Secondly, the human species is deluded into believing that the non-human animals it shares the planet with are fundamentally different to humans, when in reality they are not. Thirdly, the human species is deluded into believing that it has a fairly good understanding of the nature of the Universe, the world that surrounds it, when in reality it does not. Our immediate concern is to explore the first of these three strands; we will return to the other two strands in due course.

The technological explosion brings forth immense suffering for both human and non-human life-forms, it entails colossal planetary transformation which negatively impacts on both human and non-human life-forms, and it entails serious destabilisation of planetary biogeochemical cycles which threatens the continued flourishing of life. In the midst of these phenomena it is exceptionally hard for the human species to see that it is actually the precious saviour of life on Earth.

In order to escape this delusion one needs to come to appreciate that life on Earth's 'first line of defence' is close to being overwhelmed, and that the future thriving of life on Earth requires it to be supplanted with its 'main line of defence'. Furthermore, one needs to appreciate that the bringer forth of the 'main line of defence' is life on Earth's precious saviour, the master modifier that is the human species.

Having recognised this theoretical potential escape route it is important to realise that this delusion cannot easily be escaped from within the epoch of the technological explosion. For, within the epoch of the technological explosion the human species needs to be deluded concerning its cosmic purpose, because without such a delusion it wouldn't be able to fulfil its cosmic purpose. If the human species was not deluded, if humans en masse could see things as they truly are, then they could not possibly bring themselves to be the initiators of the immense suffering and large destabilisations that are necessary parts of the age of the technological explosion.

Our relationship to the non-human life-forms of the Earth

Let us now focus on the second strand of the state of human delusion. This is the very widespread belief that humans are fundamentally different to the non-human life-forms of the Earth. In order to power the technological explosion the human species has needed to see everything in its surroundings, including its fellow planetary life-forms, as resources for its use. The advancement in human knowledge, and the advancement in human technological prowess, has been immensely bolstered by the use of non-human life-forms in a plethora of ways. As human culture has powered forwards these life-forms have been dissected, observed, vivisected, ridden into war, put to work in agriculture, and put to the sword as a major food source to power the activities of an escalating human population.

The human species has had to justify to itself this mass exploitation of its fellow planetary life-forms. In other words, the human species has had to believe that there is a chasm between itself and its fellow planetary life-forms; it has had to tell itself that these life-forms are inferior, that they are lacking important attributes that humans possess. You will be familiar with many variants of this attempted justification. They are just animals:

They cannot think.

They are not conscious.

They lack self-awareness.

They don't have emotions.

They don't have culture.

They cannot feel pain.

They don't have a mind.

They lack language.

They cannot rationalise.

They lack feelings and are just machines.

They lack any moral worth.

They cannot use tools.

They lack creativity.

They lack intelligence.

They cannot suffer.

This attempted justification for the existence of a chasm between the human species and its fellow planetary life-forms is clearly both immensely widespread and totally absurd. A plethora of non-human life-forms on the Earth have all of the attributes listed above. There is no 'superiority-making attribute'; there is no attribute that a human possesses which all non-human planetary life-forms lack, the possession of which makes the human fundamentally different to, and superior to, the non-human. It is just that, in the epoch of the technological explosion, the human species has had to believe that such an attribute exists in order to justify its exploitation of these life-forms, and thereby maintain its sanity in the face of the immense suffering that it is causing these life-forms.

The only thing that meaningfully distinguishes the human species from the rest of the life-forms of the Earth is that the human species occupies a special place in the unfolding of life on Earth. The human species is the master modifier of its planet which has the cosmic purpose of bringing forth the 'main line of defence' in order to save life on Earth. So, the human species is superior to all of the other life-forms of the Earth, but not because of the possession of any particular attribute which particular humans possess and particular non-humans lack.

Of course, one could, quite rightly, say that the ability to bring into being the 'main line of defence' is a superiority-making attribute that the human species has, which the non-human species of the Earth lack. However, one needs to keep in mind that such an ability doesn't represent a fundamental chasm in the attributes possessed by individual humans on the one hand and individual non-human life-forms on the other hand, of the sort which has been hypothesised by the human species in order to justify the mass exploitation of its fellow planetary life-forms. The various justifications for the human utilisation of non-human animals, the hypothesised chasm, is simply a necessary delusion in the epoch of the technological explosion.

If humans were to realise the extent of the intelligence, awareness, suffering and compassion that exists in the life-forms that they were exploiting, then they couldn't possibly bring themselves to engage in such exploitative acts. Whilst the human species needs to be deluded in the epoch of the technological explosion, it will come to fully appreciate the true capabilities and attributes of its fellow planetary life-forms when our Solar System has unfolded to the point that is 'the day of the realisation of cosmic purpose'.

Our cosmic purpose

The purpose of the human species is to bring forth life on Earth's 'main line of defence'. We can refer to the 'main line of defence' as 'the sufficient deployment of technology' because when technology has complexified to the point that enables it to be deployed to create the 'main line of defence' then the future of life on Earth will be secure.

The human species is currently working tirelessly, whether it realises it or not, to bring forth the wonderful technological fruit that is 'the sufficient deployment of technology'. Humans are forever bringing forth increasingly complex technologies, forever learning more and more about the nature of the Earth and the constituents that make up the Universe, forever utilising technologies, forever fulfilling their potential and pursuing their passions; through all of these things human culture is irresistibly powered forwards.

This tireless work will ultimately lead to the bringing forth of 'the sufficient deployment of technology'. There are other technological fruits that the human species is bringing forth to protect life on Earth, such as the ability to deflect massive asteroids which are on a collision course with the Earth. However, whilst such fruits are immensely tasty side-dishes, the prize fruit, the main course, is the desperately needed 'main line of defence' – 'the sufficient deployment of technology'.

If the human species was lazy, if the vast majority of humans spent their days lounging around and not working, not learning, not enquiring, not seeking to fulfil their potential and better themselves, then life on Earth would be doomed. However, such laziness would be a ghastly denial of our nature. We simply couldn't be that lazy; such laziness is not us! We are the master modifier of our Solar System. Our nature is to master modify. We cannot not do this. We have to work tirelessly until 'the sufficient deployment of technology' has been brought into being and life on Earth has thereby been saved.

'The day of the realisation of cosmic purpose'

In the future there will come a time when knowledge of the nature of human cosmic purpose will pervade the human species. When 'the sufficient deployment of technology' has come into being, when the human species has fulfilled its cosmic purpose, it will then be possible for the human species to come face-to-face with its purpose.

On 'the day of the realisation of cosmic purpose' the nature of human cosmic purpose, the reason why humans have been working so tirelessly for so long to enable the bringing forth of increasingly complex technologies, becomes crystal clear for all to see.

On 'the day of the realisation of cosmic purpose' the human species joyously celebrates its achievements in saving life on Earth through bringing forth 'the sufficient deployment of technology'. On this day the human species comes to fully appreciate the unfolding journey of our Solar System as it has striven to bring forth life and then to maintain life. This is a wonderful journey that involves the gallant efforts of the 'first line of defence' and its inevitable supplantation by the 'main line of defence' that is magnificently brought into being through the tireless work of the human species.

Human-induced global warming

In the midst of the epoch of the technological explosion human-induced global warming is inevitably seen as a wholly bad thing; this is because there is no widespread realisation that it is an extremely welcome partial reversal in life on Earth's 'first line of defence' as it brings forth its desperately needed 'main line of defence'.

Human-induced global warming is not a wholly good thing; it can lead to undesirable outcomes, to death and suffering, in particular locations. These undesirable outcomes are part of the price that needs to be paid for the bringing forth of the 'main line of defence'.

Minimising suffering

If we were to fully accept our nature, as master modifiers, we would then realise that reining in our activities, having a minimal impact on the Earth, is not possible in the epoch of the technological explosion.

If we were to fully accept the nature of our cosmic purpose, to bring forth 'the sufficient deployment of technology', if we could see that this deployment was badly needed, and required for the future thriving, and ultimately the future survival, of life on Earth, then we could push forward at speed and bring forth 'the sufficient deployment of technology' as soon as possible.

If we were to accept our nature, and acknowledge our cosmic purpose, then we could minimise the suffering of human and non-human life-forms in the future by positively embracing our destiny. We could positively embrace the challenge of bringing forth 'the sufficient deployment of technology', and we could seek to bring it into being with the utmost haste.

The bringing into being of 'the sufficient deployment of technology' means that there will be no global warming, there will be no increase in the Earth's atmospheric temperature, so there will be no suffering to human and non-human life-forms resulting from global warming.

The longer that we are deluded, the longer that we deny our nature, the longer that we resist our destiny, the longer that we take to fulfil our cosmic purpose, the greater will be the amount of global warming that occurs, and, consequently, the greater will be the amount of avoidable suffering and death that emanates from global warming.

The folly of focusing on fossil fuel reduction efforts

Those who want us to deny our nature, those who are blinded to our cosmic purpose, believe that the phenomenon of global warming can be addressed through reducing, or ceasing, our use of fossil fuels. Consequently, they urge us to immediately significantly reduce, and/or to speedily eliminate, fossil fuel use.

Why do so many people believe this? Why do they urge us to take this course of action? The reason is that these people believe that global warming is primarily a single factor scientific issue. What does this mean? Scientists have realised that increased greenhouse gas concentrations in the Earth's atmosphere lead to an enhanced 'greenhouse effect' and cause global warming. To take a single factor scientific perspective is to jump directly from this realisation to the conclusion that the appropriate human response, the way to stop global warming, is to focus on fossil fuel reduction efforts, to focus on stopping putting greenhouse gases into the Earth's atmosphere.

We are slowly coming to appreciate the folly of this simplistic approach. We can see that life on Earth needs us to bring forth its 'main line of defence'. It is our nature to act in such a way that enables this to happen. Acting in accordance with our nature leads to the technological explosion, which entails a partial reversal in life on Earth's 'first line of defence', through the release of the previously stored fossil fuels, and leads ultimately to the fulfilment of our cosmic purpose – the bringing into being of life on Earth's 'main line of defence'.

Our mission is to come to fully appreciate why the appropriate human response to stop global warming is to bring forth the 'main line of defence' as soon as possible. Furthermore, we need to understand why any attempt to eschew the 'main line of defence' is a harmful waste of time and energy which will increase the suffering of both human and non-human life-forms.

As we are in the midst of the 'guilt-tinged' age of the technological explosion, it is understandable that there is a widespread belief that we should be focused on fossil fuel reduction efforts. However, such efforts are a great distraction which could lead to the obliteration of life on Earth. We don't have time to waste, time to be distracted, life is in peril and we need to bring forth 'the sufficient deployment of technology' as a matter of urgency.

The global warming situation that life on Earth currently faces is a potent combination of two forces: non-human-induced global warming (the incoming Solar Radiation which has almost overwhelmed the 'first line of defence') and human-induced global warming (the release of the fruits of the 'first line of defence'). The combination of forces, factors and time-lag effects currently in play relating to these two phenomena means that we need to fulfil our cosmic purpose with extreme haste.

Focusing on fossil fuel reduction efforts is an attempt to renounce our nature; it is an attempt to eschew our cosmic purpose. The longer that the human species focuses on such efforts, rather than focusing on bringing into being 'the sufficient deployment of technology', the greater is the probability that life on Earth will be obliterated. Such is the folly of focusing on fossil fuel reduction efforts.

Human-induced global warming is not a single factor scientific issue

It is of immense importance to fully appreciate that human-induced global warming is not a single factor scientific issue. It is all too easy to believe that it is. It is true that scientists have realised that increased greenhouse gas concentrations in the Earth's atmosphere lead to an enhanced 'greenhouse effect' and thereby are a force for human-induced global warming. However, this realisation does not directly lead to an indubitable conclusion considering the appropriate human response. It is not the case that one can jump straight to the conclusion that the appropriate human response is to focus on reducing fossil fuel use.

There are many other factors in play which determine the appropriate human response. Some of these factors are scientific in nature, other of these factors are non-scientific in nature. Yet, in reality the scientific and the non-scientific meld together in intricate ways, because a simple scientific finding can be interpreted in diverse ways when it is placed in broader contexts.

Two types of global warming

It is of the utmost importance to acknowledge, when considering the appropriate human response to the phenomenon of human-induced global warming, that there are actually two types of global warming. When one has acknowledged that there are two different types of global warming – the force for non-human-induced global warming (increasing Solar Radiation) and the force for human-induced global warming (the release of the fruits of the 'first line of defence') – then one immediately has a broader context within which to consider the appropriate human response.

Such an acknowledgement means that one can appreciate that life on Earth has responded to non-human-induced global warming through its 'first line of defence' which has created what we call fossil fuels. One can appreciate that the 'first line of defence' cannot continue to save life on Earth because the reduction in atmospheric carbon dioxide concentrations from 5000ppm to 280ppm means that its power, its effectiveness, has ebbed away. The 'first line of defence' is fast approaching eternal impotence in terms of its ability to offset forever increasing levels of incoming Solar Radiation. The next step is to appreciate that life on Earth needed to complexify to the point that is the human species, and the technological explosion, if it was to continue to survive and thrive. When one appreciates this, then one can see that the human release of the fruits of the 'first line of defence' – fossil fuels – is a partial reversal in life on Earth's 'first line of defence' as it brings forth its 'main line of defence'. One can also appreciate that the 'main line of defence' is 'the sufficient deployment of technology' – technology which deflects incoming Solar Radiation, thereby preventing it from reaching the Earth's atmosphere. Finally, one can come to see that such a bringing forth is the cosmic purpose of the human species.

It is not helpful to try and conceptually disentangle human-induced global warming from non-human-induced global warming. It is not helpful to believe that human-induced global warming is a single factor scientific issue that can be adequately addressed through fossil fuel reduction efforts, whereas non-human-induced global warming is a separate issue which operates over a vastly longer timescale. Timescales are irrelevant. What is relevant, what is important, is to appreciate the forces that exist in the present moment; the question of how long it took for these forces to come into existence is irrelevant.

The force for non-human-induced global warming is exceptionally powerful at the present moment, and it is interweaving with the force for human-induced global warming to create an immensely potent force. This immense force could very easily lead to the obliteration of life on Earth if the human species does not embrace its cosmic purpose.

One can appreciate that the conclusion that we have just reached is an inevitable melding of the scientific and the non-scientific. The knowledge that we have concerning the existence of non-human-induced global warming is scientific knowledge. The knowledge that we have concerning the existence of human-induced global warming is scientific knowledge. However, by themselves these two pieces of knowledge are not very fruitful. These pieces of knowledge need to be weaved together into a broad context in order to reveal the way that our Solar System, and the Earth, is unfolding. When one can see this context, then one can clearly see the nature of human cosmic purpose and how this relates to the phenomenon of global warming.

The 'non-pent-up force' for human-induced global warming

In order to appreciate the situation that we are currently facing one needs to clearly see that there are two forces for human-induced global warming – the 'non-pent-up force' which we will consider here, and the 'pent-up force' which we will consider shortly.

The 'non-pent-up-force' has two elements. The first element arises due to the fact that when the human species moves the fruits of the 'first line of defence' from underground storage to the Earth's biosphere, a small proportion of these released fossil fuels move to a new home in the Earth's atmosphere. In other words, there is a partial reversal in life on Earth's 'first line of defence' which causes atmospheric carbon dioxide levels to start to nudge upwards. This nudging upwards is a 'non-pent-up force' for human-induced global warming.

The second element arises due to the way that the human species, in the epoch of the technological explosion, has fundamentally modified the biosphere of the Earth in ways that make it harder for life on Earth to keep its atmospheric temperature down. In other words, it represents the ways that the activities of the human species have further weakened the effectiveness of life on Earth's rapidly deteriorating 'first line of defence'.

In terms of this second element, we are primarily talking about deforestation. We are talking about the intentional transformation of wonderful precious forests, which are the heart of the 'first line of defence', to habitats which suit the human species, such as cities / concrete jungles and barren expanses that are farmed by mechanised agriculture. Forests are the precious conduit between atmospheric carbon dioxide drawdown and fossil fuel creation. As far as the 'first line of defence' is concerned, concrete jungles and agricultural barren expanses are a total waste of space!

In the atrocious epoch that is the age of the technological explosion the human population size explodes, and this exploding population increasingly masters the art of intricately transforming the biosphere of the Earth to suit its needs, through the creation of such things as chainsaws, combine harvesters, trains, skyscrapers, airplanes, cars, motorways, slaughter houses and the mass utilisation of non-human life-forms as agricultural commodities. This mass transformation of the composition of the Earth's biosphere means that when the fruits of the 'first line of defence' are released into the biosphere that atmospheric carbon dioxide concentrations become higher than would otherwise have been the case. In other words, the human transformation of the Earth's biosphere is part of the 'non-pent-up force' for human-induced global warming.

All of this means that in the epoch of the technological explosion a massive 'non-pent-up force' for human-induced global warming is created through the two elements that we have considered – the movement of fossil fuels from underground storage to their new home in the Earth's atmosphere, which is exacerbated by the wholesale human transformation of the Earth's biosphere from lush forests to lifeless concrete and barren expanses that are farmed by mass agriculture.

It is crucially important to ponder on the fact that in the epoch of the technological explosion, as life on Earth brings forth its 'main line of defence', which it needs to replace its rapidly deteriorating 'first line of defence', that the 'first line of defence' becomes even weaker; this is due to the biospheric transformation initiated by the human species. This increasing weakness means that we are currently living in a period of immense danger. We need to further appreciate exactly why this is so; we need to come to appreciate all of the forces that are contributing to this immense danger; we need to behold the bigger picture, the overall situation, which encompasses our place in the unfolding of our planet, and the unfolding of our Solar System. In short, we need to urgently move past the extremely simplistic view of human-induced global warming as a single factor scientific issue.

The 'pent-up force' for human-induced global warming

We are coming to realise that human-induced global warming is not a single factor scientific issue. We need to deepen this realisation by considering the existence of the 'pent-up force' for human-induced global warming. This force has already been unleashed but it has yet to reveal its global warming impacts.

When a master modifier, in the epoch of the technological explosion, releases the fruits of its planet's 'first line of defence' – fossil fuels – the overwhelming majority of these fruits do not move into a new home in the atmosphere of that planet; if they did they would cause very significant immediate global warming. Rather, what happens is that a colossal force for future global warming is created, and, worryingly, it is easy to be blinded to the existence of this force. We need to fully appreciate that this time-lag 'pent-up force' for future global warming exists. If we are blinded to it, it could be curtains for the future of life on Earth.

Once released, the overwhelming majority of released carbon gets sucked straight into the planet's ocean where it sets off on a lengthy journey through what, on the Earth, we have called the 'thermohaline circulation'. The thermohaline circulation involves dense cold water sinking at high latitudes and travelling very slowly through the ocean depths until it eventually reaches the northern Indian Ocean and the northern Pacific Ocean where it resurfaces. After the released carbon sinks into the thermohaline it takes from 100 years to 1000 years to emerge into the atmosphere. This means that almost all of the carbon that has become temporarily stored in the thermohaline circulation since the start of the Industrial Revolution has yet to emerge into the atmosphere. Oh no!!! When the bulk of this carbon starts to be released it is set to turn from a 'pent-up force' for human-induced global warming to a bringer of a large increase in the temperature of the Earth's atmosphere.

Worryingly, the world to which this carbon is returning is a very different world to the one that it left; it will be released into a human-transformed biosphere of concrete jungles and agricultural barren expanses. This transformed biosphere massively exacerbates the magnitude of the rise in atmospheric temperature that is set to occur. So, the human transformation of the biosphere, in the age of the technological explosion, is part of both the 'non-pent-up force' for human-induced global warming and the 'pent-up force' for human-induced global warming.

If this large increase in atmospheric temperature was to occur, it is likely to trigger other large-scale discontinuities such as a runaway 'greenhouse effect' resulting from the destabilisation of methane clathrate reservoirs. By around the year 3000 – when all of the carbon that is currently stored in the thermohaline has been released – the atmospheric temperature of the Earth could very easily be way too hot for the existence of the human species, and for any other complex life-forms.

Furthermore, if this large increase in atmospheric temperature was to occur, life on Earth would be forever doomed because its 'first line of defence' would have been fatally impaled, and its attempt to attain its 'main line of defence' in time would have failed. The Earth's biosphere would have transitioned into a new hotter state and the future would only ever involve increasingly hot atmospheric temperatures due to the forever increasing strength of the force for non-human-induced global warming. The Earth would never again be able to support the existence of complex life-forms. Of course, as we are considering a 'pent-up force', this would be the case even if humans were to stop using fossil fuels today.

Let us repeat the point, because it is a crucial one. We are considering the future impacts of a force that has already been created, a 'pent-up force', so it matters not a jot whether or not humans reduce their future fossil fuel emissions, or even eliminate them completely. Such efforts are just a great distraction. If such efforts distract us sufficiently they will be the cause of the obliteration of life on Earth!!! What a travesty that would be.

Acknowledging the existence of this 'pent-up force' for human-induced global warming is essential. The existence of this force means that fossil fuel reduction efforts are not a solution to the human-induced global warming situation that we are currently facing, let alone the overall global warming situation that we are currently facing (the combined force resulting from the force for non-human-induced global warming, and the two forces for human-induced global warming – the 'non-pent-up force' and the 'pent-up force').

We can increasingly appreciate that human-induced global warming is not a single factor scientific issue.

We can increasingly appreciate the folly of focusing on fossil fuel reduction efforts.

We need to immediately get our heads out of the sand and start to act in an appropriate way in order to save life on Earth! Let us bring forth 'the sufficient deployment of technology' as soon as possible!

The timing of the bringing forth of the master modifier

I have been painting a picture of our unfolding Solar System. This picture involves the human species, the master modifier, the saviour of life on Earth, the bringer into being of the 'main line of defence', coming into existence at the exact same time as the Earth's 'first line of defence' is frail and deteriorating, and close to collapse. Is this timing a happy coincidence?

Obviously, if our Solar System unfolded to the point at which the 'first line of defence' collapsed before the master modifier came into being, then life on Earth would be decimated; the Earth's atmospheric temperature would forever be too hot for the thriving of life.

Alternatively, it is possible to imagine that at the point of Solar-Systic unfoldment that the master modifier came into being that the 'first line of defence' was still robust, in rude health and in its relative youth. In this case, there would be no urgency to bring forth the 'main line of defence'.

It is obviously in the interests of life on a planet to bring into being its master modifier as soon as it possibly can, because this significantly boosts its survival chances. Yet, this feat is no easy task. When life first evolved on the Earth it was extremely simple in its composition, which means that it had a very limited ability to intricately mould its surroundings. Life on Earth has gone through a process of complexification from this extremely simple composition to the exceptionally complex arrangement that is the master modifier. The master modifier is able to mould its surroundings in the most amazingly delicate and intricate ways. The master modifier is the endpoint of an exceptionally long journey which has involved life on Earth ever so gradually increasing its ability to modify its surroundings in increasingly delicate and intricate ways. So, whilst the ideal scenario for life on Earth is to bring forth its master modifier as soon as life arose on the planet, this was obviously an endeavour that was going to take an exceptionally long period of time.

Whilst it takes life on a planet an exceptionally long stretch of time to bring into being its master modifier, the 'first line of defence' of life on a planet also lasts for an exceptionally long stretch of time. This means that it is easy to comprehend why there is likely to be a similarity in timing between the bringing into being of life on Earth's master modifier and the weakening and deterioration of life on Earth's 'first line of defence'. This similarity in timing is certainly not a cause of great surprise. Perhaps this similarity in timing reveals a deeper order at work in our Solar System, within which things are brought into being when they are needed, and things work out for the best.

What does this mean for our current situation? We can come to fully appreciate that we have inevitably been bought forth into existence at the time that life on Earth's 'first line of defence' is extremely frail and badly deteriorating. We can accept that our purpose is to bring forth 'the sufficient deployment of technology' to save life on Earth and give it a long and glorious future. We can realise that achieving our cosmic purpose involves us utilising the abilities that our Solar System has provided us with, thereby resulting in a technological explosion. We can come to see that this involves us causing immense disruption to the Earth, including releasing the fossil fuels which are the stored fruits resulting from the functioning of life on Earth's 'first line of defence'. We can come to appreciate that our focus of concern should not be human-induced global warming. Our focus of concern should be the overall situation that we face, given all of the above.

Précis

We have been coming to appreciate that human-induced global warming is not a single factor scientific issue – one cannot jump directly from the scientific appreciation that humans are enhancing the 'greenhouse effect' through the release of fossil fuels, to the conclusion that the appropriate response is to reduce and eliminate fossil fuel use. Rather, the phenomenon of human-induced global warming needs to be seen against the backdrop of the unfolding of our Solar System, as one factor which is intertwining with many other factors and forces to give rise to the overall situation that we currently face.

We have been exploring this overall situation, this broader context within which the phenomenon of human-induced global warming is situated, through considering three intertwining forces – the force for non-human-induced global warming, the 'non-pent-up force' for human-induced global warming, and the 'pent-up force' for human-induced global warming. We also need to consider another extremely powerful force that is in play – 'the force that is human nature'.

'The force that is human nature'

In order to more fully appreciate the situation that we are currently in, and consequently the course of action that we need to take, we need to reflect on our own nature.

We are not alien to our fellow planetary life-forms, alien to our planet, alien to our Solar System. It is not as if we travelled from an alternate universe and plonked ourselves on the Earth. Rather, we are a fundamental expression of the nature of the Universe. The Universe, our Solar System, our planet, our fellow life-forms, they all brought us into being; they evolved us. We are an outgrowth of all that is. Our abilities, our talents, our skills, our feelings, they are all fundamental expressions of the underlying nature of the Universe, our Solar System, our planet, and the life which brought us into being.

Furthermore, we didn't evolve as a fluke. Every life-form, and every part of the Universe, modifies its surroundings to the best of its ability, using the talents/skills that are at its disposal. This obviously applies to us. Indeed, we are the zenith of the ability to modify; we are the master modifier of our Solar System. Our skills, our abilities, our talents, and our passions, are gifts that have been endowed to us by our Solar System, and it is our nature to utilise these things as best we can. We didn't create these things, let us repeat, they are gifts from our Solar System. If we use these gifts, that cannot be a bad thing; we weren't given these gifts to cause mischief! We were brought into being in order to master modify; we were brought into being so that we could make use of the gifts which we have been provided with.

The natural/normal state of affairs is for a human to seek to make use of their abilities, skills, talents and passions. Some humans are gifted scientists/engineers/inventors. Some humans are gifted actors. Some humans are gifted entrepreneurs / business people / politicians. Some humans are gifted teachers/writers/academics. Some humans are gifted illustrators/mathematicians/musicians.

Some humans are gifted nurses/doctors/surgeons. Some humans are gifted footballers/athletes. Some humans are gifted spiritual leaders / comedians / artists. Some humans are gifted at fossil fuel extraction. Some humans are gifted environmentalists. We need to be utilising all of our gifts.

It is, of course, possible for a particular human to have a range of abilities/skills/talents/passions that spans all of the above domains; however, this is not normal. The vast majority of humans are gifted at some things and terrible at other things. Furthermore, some humans are immensely passionate about things which other humans are wholly indifferent to. It is good that these differences exist. More than this, these differences are wonderful! The human species has been provided with the diverse range of abilities/skills/talents/passions that it needs in order to fulfil its cosmic purpose and bring forth 'the sufficient deployment of technology'.

Why do humans act the way that they do? Humans seek to be happy; they seek to have a pleasant and meaningful life. How is this accomplished? A human needs to know what they are passionate about. A human also needs to know what their particular abilities/skills/talents are. It is highly likely that there will be a significant crossover between the two: there will be lots of ways in which a human can deploy their abilities/skills/talents in accordance with their passions. If such a deployment occurs then we can say that a human is fulfilling their potential. To talk of human nature is to talk of the plethora of human passions and their deployment. In other words, to talk of human nature is to talk of humans fulfilling their potential.

The activity of individual humans as they seek to fulfil their potential is in itself a force. When we consider the activities of all the humans that exist we can start to get a grasp of how this force operates at the planetary level, across the Earth. The cumulative day-to-day activity of every human across the planet generates 'the force that is human nature'.

In the age of the technological explosion 'the force that is human nature' is dominated by the master modification activities of the human species. In this age, the natural state for the overwhelming majority of humans is to act so as to fulfil their potential through engaging in immense modifications of their surroundings. This could be through driving their car a lot, partaking in airplane flights, using a combine harvester, playing the piano, operating a chainsaw, mowing their lawn, talking on their mobile phone, buying/consuming/hoarding an immense plethora of things, and bringing forth or utilising a whole range of technological devices. This colossal force, this unquenchable thirst for master modification, is unstoppable in the epoch of the technological explosion.

The unstoppable nature of this force in the epoch of the technological explosion is something to be celebrated, not something to be feared. For sure, it leads to planetary transformation, it involves mass deforestation, it entails the release of stored fossil fuels, it triggers the destabilisation of biogeochemical cycles, and it generates immense suffering for both human and non-human life-forms. However, this unquenchable thirst, this colossal force, this exuberant expression of the fundamental nature of the Universe, ultimately leads to the bringing into being of life on Earth's desperately needed 'main line of defence' - 'the sufficient deployment of technology'. In the absence of this force, life on Earth would be doomed. Whilst the epoch of the technological explosion is brutal for the particular life-forms that exist within it, it is simultaneously an immensely precious epoch for life on Earth.

'The force that is human nature' needs to be acknowledged, accepted and embraced. It represents our fundamental nature, it motivates us, it flows through us, and its continued unleashing is required if life on Earth is to have a long and glorious future. There is no point seeking to repudiate this force, or wish it away; such an endeavour will do much more harm than good. The master modifier has to master modify!

The concoction of forces that inevitably threaten life on Earth at precisely this moment

Our focus of concern needs to be the delicate interplay of forces that are currently in play. There are a concoction of forces that are inevitably currently combining at precisely this moment in the evolution of our planet and Solar System. We are living at the moment of extreme danger for life on Earth.

The force for non-human-induced global warming is currently exceptionally powerful and has almost overwhelmed life on Earth's 'first line of defence'. The extreme potency of this force is inevitably bolstered, in deeply worrying ways, by the activities of the master modifier as it has brought forth life on Earth's technological explosion. The master modifier has released the fruits of the 'first line of defence' and has thereby unleashed a potent, and potentially devastating, 'pent-up force' for future human-induced global warming, which will be unleashed on the Earth's atmosphere when the slow-moving carbon starts to gush out of the thermohaline circulation. The activities of the master modifier have also severely weakened life on Earth's already extremely frail 'first line of defence', its ability to regulate its atmospheric temperature, through the transformation of wonderful forests into desolate concrete jungles and barren expanses that are farmed by mechanised agriculture. This biospheric transformation vastly exacerbates the immense threat to life on Earth which will exist when the slow-moving oceanic carbon reaches its atmospheric destination.

The forces that we have just considered form an exceptionally potent concoction; they pose an immense threat to the continued flourishing of life on Earth. This concoction is about to get even more potent! We need to add to it 'the force that is human nature'. As we are in the midst of the age of the technological explosion 'the force that is human nature' is inevitably propelling us down the path of increasing future master modification as humans continue to fulfil their potential. This means that the immediate future entails a greater human impact on the planet, greater perturbations, greater destabilisations, greater transformations, and greater danger in the present moment.

We need to acknowledge this perilous concoction of forces. We need to embrace our destiny. Life on Earth needs us to embrace our cosmic purpose. The time of danger is now. The time for action is now. Will we succeed or will life on Earth perish?

Tinkering with the 'greenhouse effect'

Most people are wholly unaware of the concoction of extremely dangerous forces that are currently threatening the continued flourishing of life on Earth. Most people see global warming as a single factor scientific issue and hence believe that a 'small scale' response, or a series of such responses, is an appropriate response to the situation that we currently face. For example:

Let's plant lots of trees!

Let's try and cut our fossil fuel emissions!

Let's try and reduce the number of airplane flights!

Let's encourage people to buy electric cars!

Let's sell our shares in fossil fuel companies!

Let's invest in renewable energy!

Let's try and recycle more and buy fewer things!

Let's try and persuade the government to decarbonise the economy!

If you are aware of the perilous concoction of forces that currently exist, then you will realise that all of this is totally hopeless and tragically woefully inadequate. These 'small scale' responses are really a type of 'return to the past' thinking. This goes something like: we have messed things up through our destabilising activities, so we now need to try and rein our activities in, to 'leave things to nature', so that we can restore the balance that existed in the past. What we desperately need is a 'big scale' response; we need to embrace our nature as the master modifier of our planet; we need to push forwards, not backwards; we need to increase our presence on the planet, not rein it in. If we fail to do this we will, through our inaction, be cruelly condemning, and ultimately extinguishing, the life that has arisen on the Earth.

All 'small scale' responses involve attempting to tinker with the 'greenhouse effect'. Such tinkerings don't go anywhere close to the scale of the response that is needed to deal with the concoction of forces that are aligned against us at this moment in time. As you are well aware, the reduction in atmospheric carbon dioxide concentrations to 280ppm at the start of the Industrial Revolution means that the 'first line of defence' has almost worn out. On top of this we have inevitably unleashed both the 'non-pent-up force' for human-induced global warming, and the 'pent-up force' for human-induced global warming, so the situation is now extremely dire.

We urgently need to stop thinking about tinkering with the 'greenhouse effect' and instead focus on bringing forth 'the sufficient deployment of technology' as soon as possible. However, whilst we are bringing forth 'the sufficient deployment of technology' we might need to deploy our master modification abilities to buy a little more time not through a tinkering, but through a major manipulation of the 'greenhouse effect' which entails initiating a massive drawdown of carbon from the atmosphere. This would probably entail the creation of ocean seaweed forests. However, such a massive drawdown would be nothing more than a less than ideal temporary patching up of the fatally impaled 'first line of defence'. It is not a long-term solution and it could have unwelcome side-effects; yet, it might be needed if we cannot bring 'the sufficient deployment of technology' into existence in time. What is really needed is not any kind of manipulation of the 'greenhouse effect'; what is needed is the 'main line of defence' itself – 'the sufficient deployment of technology'.

Tinkering with the 'greenhouse effect' will not give life on Earth a long and glorious future; it won't even give life on Earth a short and glorious future. What life on Earth urgently needs is its most powerful weapon – 'the sufficient deployment of technology'. Let's get to work!

The next 300 years

Given that the amount of fossil fuels burned by humans has been enormously high in the past 100 years, and was also extremely high in the preceding 100 years, and given that the overwhelming majority of the carbon thereby released was sucked into the thermohaline circulation, where it embarks on an oceanic journey ranging from 100 years to 1000 years, after which it emerges into the atmosphere, you can quickly do the maths and work out that the next few hundred years are going to be interesting!

The next 300 years are the time of real danger for life on Earth. This is the time when the force for human-induced global warming is at the zenith of its potency; this is mainly due to the 'pent-up force' coming to fruition, but in this period the 'non-pent-up force' also reaches its peak strength as the epoch of the technological explosion reaches its zenith. And, of course, the force for non-human-induced global warming is already exceptionally potent, having almost overwhelmed life on Earth's 'first line of defence'. The combination of these extremely powerful forces, the perilous concoction that will reach its maximum potency in the next 300 years, is something that we should be extremely concerned about.

We are the glorious saviours of life on Earth, but in order to fulfil our cosmic purpose we have had to be deluded; we have had to bring forth the brutal epoch of the technological explosion; we have had to bring forth immense suffering both for ourselves and for our fellow planetary life-forms; we have had to perturb the Earth's biogeochemical cycles in a way that exacerbates the risk that life on Earth will be decimated in the present moment. This has all been necessary so that we can bring forth 'the sufficient deployment of technology' in order to save life on Earth, giving it a long and glorious future.

We need to immediately get our collective heads out of the sand and start to act in an appropriate way in order to save life on Earth! Let us free ourselves of the erroneous belief that the situation that we currently face can be resolved through fossil fuel reduction efforts. We are living at the pivotal moment in the unfolding of life on Earth, if we continue to focus on fossil fuel reduction efforts, and ignore 'the sufficient deployment of technology', then life on Earth will be decimated in the near future. There is no time to waste. Let us fulfil our destiny. Let us fulfil our cosmic purpose. Let us bring forth 'the sufficient deployment of technology' as soon as possible!

The sufficient deployment of technology'

We have seen that if life on Earth is to survive and have a long and glorious future then it needs its frail and deteriorating 'first line of defence' to be replaced by its 'main line of defence'. In other words, it needs its master modifier, the human species, to bring forth 'the sufficient deployment of technology'. The 'main line of defence' entails the deployment of technology to deflect incoming Solar Radiation, thereby preventing it from reaching the Earth's atmosphere. It is our cosmic purpose to bring forth 'the sufficient deployment of technology' and the sooner that we do this the better. The aim of this book is to highlight the need for 'the sufficient deployment of technology' and to explain why such a deployment is not only a necessity, but also an exceptionally precious and wonderful thing; it is not something to be afraid of, or something to be resisted.

What, exactly, does 'the sufficient deployment of technology' look like? What does it physically entail? These are questions for the engineers, for the experts in technology; they are not questions for me. As I have just said, the aim of this book is to make clear the need for 'the sufficient deployment of technology', so that the engineers and policymakers can be encouraged to push full steam ahead and bring it into being.

Any technology that can deflect incoming Solar Radiation to the Earth in a reliable and controlled way, thereby actively regulating the Earth's atmospheric temperature to keep it down to a level that is favourable for the continued thriving of life, could be 'the sufficient deployment of technology'.

Let us briefly consider a few possible physical instantiations of 'the sufficient deployment of technology'. A space sunshade is the most obvious candidate. The sunshade would seemingly need to be placed at the L1 Lagrangian point between the Earth and the Sun, which is located 1.5 million kilometres above the Earth. At Lagrangian points a smaller object will maintain its position in relation to two larger bodies that are in orbit. The sunshade at the L1 point could either be a single large object or a plethora of small objects.

In terms of the former possibility, in 1989 James Early designed a 2000 kilometre glass occulter lens which could be placed at the L_1 point. This lens would disperse Solar Radiation away from the Earth's atmosphere. In a similar vein, in 2004 Gregory Benford calculated that a concave rotating Fresnel lens placed at the L_1 point, which was 1000 kilometres across, and a few millimetres thick, would deflect up to 1% of incoming Solar Radiation to the Earth's atmosphere. Rather than deploying a lens, in 1997 Edward Teller, Lowell Wood and Roderick Hyde proposed that a 3000 ton thin wire mesh diffraction grating could be placed at the L_1 point.

In terms of the latter possibility, in 2006 Roger Angel's design involved placing 16 trillion small disks at the L_1 point. Each disk needs to have a 0.6 metre diameter and a thickness of 5 micrometres. This number and design of disks would deflect 2% of the incoming Solar Radiation to the Earth.

If we zoom forwards in time, if we imagine the state of our Solar System in a few thousand years, we can be confident that there will be technological devices at the L_1 point which deflect Solar Radiation away from the Earth's atmosphere. In the immediate future, while we are still escalating our technological expertise, there is also the possibility for using technology to prevent Solar Radiation from reaching the lower atmosphere of the Earth (the troposphere) through manipulating the composition of the Earth's upper atmosphere (the stratosphere). For example, aircraft could be used to put sulfate aerosols into the stratosphere where they would reflect incoming Solar Radiation back into space.

Could such a manipulation of the stratosphere be a long-term solution? Could such a manipulation be 'the sufficient deployment of technology'? It is very hard to imagine that this could be the case. Such a manipulation would bring with it several undesirable side-effects, it would be hard to attain the required levels of control and reliability, and it is a relatively short-term manipulation which requires regular replenishment. The safest and the most elegant form for 'the sufficient deployment of technology' to take is the deployment of a technological sunshade at the L_1 point.

Embracing extreme uncertainty

It is time for us to return to our consideration of the state of human delusion which exists in the age of the technological explosion. We earlier explored the first two strands of this state of delusion. The third, and final, strand of this state of delusion is that the human species believes that it has a better understanding of the nature of the Universe, and a better understanding of the unfolding of the biogeochemical cycles of the Earth, than it actually has. One consequence of this 'understanding deficit' is that the human species consistently underestimates the level of uncertainty that exists. Sometimes the level of this underestimation is ginormous!

In the realm of environmentalism, this results in a range of forecasts which the human species believes to be very diverse and broad in scope, but which are all actually in a very narrow and conservative range. For example, the IPCC has produced a range of forecasts concerning the possible level of the Earth's atmospheric temperature by the year 2100. This range of forecasts is formulated within a belief system which necessitates that the human species believes that it has more knowledge than it actually has. What this means is that even the most pessimistic forecasts could be overshot, and not just by a small margin, but by an enormous margin!

If we could fully comprehend the extent of the uncertainty that exists then this would be most advantageous. Embracing extreme uncertainty leads to the realisation that there is an extreme need for a vastly increased ability to control and modify. Such an increased ability to control and modify provides the means through which a high level of future uncertainty can be adequately dealt with by the human species for the benefit of life on Earth. In other words, embracing extreme uncertainty paves the way for the speedy bringing forth of both 'the sufficient deployment of technology' and other technological protectors of life on Earth.

Cherishing the living

We might be the master modifier, the zenith of the evolutionary progression of life on Earth, the saviour of life on Earth, the most precious life-form in our Solar System, but we don't have to see our fellow planetary life-forms as inferior, as resources to be used and eaten. We needed to see our fellow planetary life-forms this way in the past, we needed to use and abuse non-human animals in the past, but there is certainly no need to do so now. When an individual life-form dies this is a cause for sadness; when it survives this is a cause for celebration. Let us cherish each and every one of these immensely precious things.

Paradise on Earth

When we have passed through the painful era that is the epoch of the technological explosion, when we have fulfilled our cosmic purpose and brought forth 'the sufficient deployment of technology', then the Earth will transition into a new phase of its existence.

Life on Earth has a long and glorious future. The super-advanced technology that will be maintaining the atmospheric temperature in order to keep it favourable for the continued thriving of life on Earth enables this. In this new phase of Solar-Systic unfoldment every human will appreciate the nature of human cosmic purpose because 'the day of the realisation of cosmic purpose' occurs. In this new era the life-forms on the Earth will be living together more harmoniously than they ever did in the previous stages of the Earth's history.

In this new phase, the age of the spiritual explosion, the human species is living sustainably and it has the utmost respect for all of the life-forms that it shares the Earth with. Paradise will have come into existence on the Earth.

Celebration

Wouldn't it be nice if we could celebrate our existence? We are the saviours of life on Earth! The survival of our non-human brothers/sisters is our responsibility; we need to take care of them through bringing forth 'the sufficient deployment of technology'. Oh, if we could see all this, we could then celebrate our existence! Oh, if we could realise how important and special we are! Oh, if we could see the cosmic responsibility that rests on our shoulders! Could we possibly bear to come face-to-face with this immense responsibility?

Our Solar System is celebrating us; it is joyous at our arrival. Yet, we are miserable! We are in a state of despair! We inevitably tell ourselves that we are selfish and greedy and that our presence on the Earth is destructive and harmful. We tell ourselves that we are the enemy of our non-human brothers/sisters. How deceived we are. How wonderfully deceived!

One day humans will be celebrating. One day will be 'the day of the realisation of cosmic purpose'. On this day, every human will appreciate the value of all humans, all cultures, all perspectives and all life-forms. On this day, every human will see the nature of human cosmic purpose; they will see the wonderful journey of Solar-Systic unfoldment laid out before them in its entirety.

Human Purpose & the Universal Pursuit of Ecstasy

The book that you are currently reading has a companion: 'Human Purpose & the Universal Pursuit of Ecstasy' (2019). This companion book is much longer than 'Global Warming & Human Cosmic Purpose'. There are two reasons for this. Firstly, it explores in greater depth some of the themes that arise in this book. Secondly, it places these themes in a much larger context by exploring the very nature of the Universe and the way that it unfolds through time; this exploration penetrates through to the heart of the third strand of human delusion, a strand which we have barely been able to get below the surface of in this book.

The topic that we have been focused on in this book, the question of how the human species relates to the rest of life on Earth, when viewed from the perspective of the unfolding of life on Earth as it strives to survive, and how this relates to human-induced global warming and non-human-induced global warming, is an important enough topic to merit its own independent consideration.

In 'Human Purpose & the Universal Pursuit of Ecstasy' I explain why the term 'human species' is most appropriately used not as a biological term; rather, the term should be used to refer to the master modifier of its host planet and solar system. The human species = the master modifier. This is why human-induced global warming is an inevitable stage of unfoldment of any thriving life-bearing planet. As you are well aware, human-induced global warming is a sign that all is well on a planet; it is a sign that life on that planet is well on its way to bringing forth its 'main line of defence' which it needs to replace its inevitably deteriorating 'first line of defence'.

It is worth noting that there is a slightly different tone in the two books. In 'Human Purpose & the Universal Pursuit of Ecstasy' I take a relatively fatalistic approach to the way that human culture, the Earth, and our Solar System, unfolds. I say that we are cosmic puppets who are powerless to meaningfully change the way that things will unfold in the future. Whereas, in this book, I am making an urgent 'call for action', a plea to radically change our approach to the phenomenon of global warming, so that we can change course and thereby minimise the amount of suffering and death that occurs, whilst also giving life on Earth a long and glorious future. Such a plea, surely, you might think, only makes sense if we are able to change the way that things will unfold in the future in a significant way. This difference in tone is not contradictory. For, the very writing of this book, and any consequences that might flow from this event, could be part of the relatively fatalistic unfolding of our Solar System!

In this book I have sought to get straight to the point, straight to the heart of the issue of how the human species relates to the other life-forms on the Earth, and why this means that we are the saviour of life on Earth which has a cosmic purpose to urgently fulfil. I have also sought to clearly explain why the overwhelming majority of humans alive at the moment inevitably see the human species in a negative light, due to their existing in the midst of the disruptive epoch that is the 'guilt-tinged' age of the technological explosion. If you would like a deeper perspective on the nature of the Universe, and the way that our Solar System unfolds through time so as to bring forth the human species, then 'Human Purpose & the Universal Pursuit of Ecstasy' is the book for you.

Embracing our destiny / Embracing your destiny

I hope that you have come to appreciate that the future of the human species, and the future of life on Earth, will be determined by the way that we respond to the potent concoction of forces that inevitably threaten life on Earth at precisely this moment.

If we continue to be deluded, if we continue to be blinded to our cosmic purpose, if we continue to believe that we are the destroyers of life on Earth, rather than its precious saviour, then the future is extremely bleak. It is the current predominance of this belief, a fleeting predominance which is an inevitable aspect of the 'guilt-tinged' age of the technological explosion, which has led to the pandemic of urgings to 'save the planet' from the destructive and destabilising activities of the human species. In turn, this has resulted in the widespread belief that the appropriate human response to the phenomenon of human-induced global warming is to attempt to rein in our activity, to reduce our fossil fuel emissions, to reduce our impact on the Earth and to 'leave things to nature'. Such a response, as we have seen, has devastating consequences. It results in needless global warming, needless climate change caused by global warming, needless immense suffering, and needless death.

Let us escape our delusion! Let us wake up to the fact that human-induced global warming is not a single factor scientific issue! Let us appreciate the wider context in which the phenomenon of human-induced global warming is situated! Let us realise our nature! Let us embrace our destiny! Let us fulfil our cosmic purpose! Let us create a future in which there is no more needless suffering and death emanating from our delusion!

We need to increase our presence on the Earth; we need to extend the reach of our master modification abilities. We need to move full steam ahead and embrace the challenge of deploying advanced human-created technologies which regulate the Earth's atmospheric temperature. If we attempt to rein in our presence on the Earth, to 'leave things to nature', then the day of decimation for life on Earth will be nigh. If we were to rein in our presence on the Earth, we would be tragically letting down the life that has arisen on the Earth. What an utter disappointment we would be!

Life on Earth needs us!

Life on Earth needs you!

Go forth; activate your passions; utilise your abilities, skills and talents; fulfil your potential!

Go forth; throw off the shackles of your delusion!

Go forth; embrace the glorious future in which the human species saves life on Earth through the speedy bringing into existence of 'the sufficient deployment of technology'!

Our connection

If you have any questions concerning the contents of this book, if you think that we might be able to work together, or if you just fancy a chat, then feel free to contact me:

theuniversalpursuitofecstasy@gmail.com

I also have a website:

www.drcphilosophy.com

At the time of publication I am living on the outskirts of Exeter.

I very much hope that you have found this book to be of some merit.

May the life that has arisen on the Earth have a long and glorious future!

Have a fabulous life.

Neil / Dr C